Alan Araújo
Cláudia Braga

Acid rain in a virtual experiment

Alan Araújo
Cláudia Braga

Acid rain in a virtual experiment

An activity proposal for teaching chemistry using the yenka chemistry software

ScienciaScripts

Cover image: www.ingimage.com

This book is a translation from the original published under ISBN 978-3-330-76956-4.

Publisher:
Sciencia Scripts
is a trademark of
Dodo Books Indian Ocean Ltd. and OmniScriptum S.R.L publishing group

120 High Road, East Finchley, London, N2 9ED, United Kingdom
Str. Armeneasca 28/1, office 1, Chisinau MD-2012, Republic of Moldova, Europe
Managing Directors: Ieva Konstantinova, Victoria Ursu
info@omniscriptum.com

Printed at: see last page
ISBN: 978-620-8-62426-2

SUMMARY

ACKNOWLEDGMENTS

To God for giving me the strength to carry on in the most critical moments and the will to complete this phase of my life.

To my mother Elsimar Peixoto, whom I love so much.

To all my friends and colleagues at PIBID. I spent two great years with them and I don't regret anything I experienced with them.

My PIBID coordinators, Professors Karen Weber and Teresa Saldanha, for believing in me.

To my advisor, Prof. Dr. Clàudia Braga, who was patient and willing to help me carry out this work, and to those who agreed to be part of the examining board, Prof. Dr. Clàudia Cunha and Prof. Dr. Ricardo Alexandre.

To Saloana Santina, the chemistry teacher at the school where this work was carried out, for having authorized it with her full support.

I would like to thank all my real and virtual friends for the help and strength I have received and continue to receive. Thank you from the bottom of my heart!

I dedicate this work to my family, especially my mother Elsimar Peixoto, and to all my friends for their love, companionship and strength.

1 INTRODUCTION

The search for more meaningful learning has been a major challenge when it comes to teaching chemistry in Brazil. However, we must consider that many innovative teaching methods have emerged in recent years. These methods seek to move away from traditional teaching, which refers to the "passive action of the learner who is still treated as a mere listener to the information that the teacher exposes" (GUIMARÂES, 2009) and which is a notorious reality in the country's public schools.

We know that there are a number of methods that have many resources at their disposal, providing innovation and more meaningful teaching and learning. Among the many resources that can be used to help students, we can highlight the computer as a representative of the various IT tools and *software* used in education, increasingly becoming an amplifier of potential in the training and improvement of students, teachers and the educational institution itself (RIBEIRO, 2009).

As stated by Chassot (1990, *apud* ANDRADE, 2012), understanding the reasons and objectives that justify and motivate the teaching of chemistry can be achieved by abandoning classes based on the simple memorization of names and formulas and linking them to the students' day-to-day knowledge and concepts. In other words, getting them to learn about the chemistry of their everyday lives using tools that facilitate learning, such as computers.

It is known that the teaching of chemistry in most public schools is still done by memorizing formulas and isolated knowledge. This makes the subject uninteresting for the student and fails to achieve the objectives of the teaching-learning process. As an attempt to overcome this, experimental classes have emerged as another way of opening up new horizons in the innovation and renewal of new methodologies in the teaching of chemistry. However, few public schools have a science laboratory, an essential environment for practical classes. However, we know how difficult it is to build one, considering the generation of waste and the lack of qualified staff in the institution to manage it, as well as the high cost of purchasing reagents and equipment.

Unlike science laboratories, there are computer labs in most of the country's public schools - specifically in Paraiba. This is where we find the opportunity to take advantage of the resources that computers can offer by applying them to the teaching of chemistry.

Given this, why not combine the resources of a science laboratory with the computer, using *software* that allows students to observe and interact with chemical phenomena? As long as it is used appropriately, as part of a coherent and organized educational approach, there will be no reason not to use such software, which is one of the aims of this work.

It is also important to consider that the use of computers in education only makes sense to the extent that teachers see them as a tool to support their teaching activities, as an instrument for planning and carrying out interdisciplinary projects, as an element that motivates and at the same time challenges the emergence of new teaching practices, making the teaching-learning process an innovative, dynamic, participatory and interactive activity (TEXEIRA; BRANDÂO, 2003).

The software is a virtual laboratory called *Yenka Chemistry* and is compatible with many operating systems. *Yenka Chemistry* simulates a series of experiments that are carried out in the laboratory, which can make learning more meaningful. The main purpose of this work is to apply an experimental activity with the help of the program, addressing the contents of the acid rain theme, and to show that it is useful as a support tool in the teaching of chemistry, becoming a meaningful and motivating alternative in the teaching of chemistry.

This work was applied at the Presidente Mèdici State School in Joao Pessoa and showed that the virtual laboratory - *Yenka Chemistry* - promotes a better understanding of chemical phenomena, broadening the understanding of the transformations of matter, as well as being a motivating agent in teaching and learning.

2. OBJECTIVES

2.1. GENERAL OBJECTIVES

To give students an experimental lesson on acid rain, with the help of the *Yenka Chemistry software* and to present its relevance as a support tool in chemistry teaching to improve teaching and learning, as well as the challenges and difficulties encountered in taking advantage of these resources.

2.2. SPECIFIC OBJECTIVES

- Draw up a lesson plan and an experimental/virtual script contextualized to the topic of acid rain;
- Introducing and using the *Yenka Chemistry software* as a support tool for teaching chemistry;
- Integrating chemistry teaching and IT;
- Evaluate the methodology used in an opinion poll;
- Promote students' curiosity and interest in Science - Chemistry, using the virtual laboratory.

3. THEORETICAL BASIS

3.1. CHEMISTRY AND CONTEXTUALIZATION

Contextualization in teaching seeks to bring students' everyday lives into the classroom, while at the same time bringing students' everyday lives closer to scientific knowledge. Such actions, in complex subjects such as chemistry, are extremely important (ANDRADE, 2012).

Contextualizing chemistry is not about making an artificial link between knowledge and the student's everyday life. It is not to cite examples as an illustration at the end of some content, but to contextualize is to propose "real problem situations and seek the knowledge necessary to understand them and try to solve them" (BRASIL, 1999).

According to the National Curriculum Parameters for Secondary Education (PCNEM), contextualizing content in classes with students means first assuming that all knowledge involves a relationship between subject and object. In these documents, contextualization is presented as a resource which seeks to give new meaning to school knowledge, enabling students to learn more meaningfully (BRASIL, 1999).

The PCNEM also states that "the contextualized treatment of knowledge is the resource that the school has to remove the student from the condition of passive spectator" (BRASIL, 1999). It also states that "it is possible to generalize contextualization as a resource for making learning meaningful by associating it with experiences from everyday life or with knowledge acquired spontaneously" (BRASIL, 1999). These documents also state that contextualization should serve as a pedagogical resource capable of contributing to the construction of knowledge and the formation of higher intellectual capacities (BRASIL, 1999).

For Piaget (1977), knowledge is achieved through continuous and renewed constructions based on interaction with reality. It does not occur through mere copying of reality, but through assimilation and accommodation to previous structures which, in turn, create the conditions for the development of subsequent structures.

According to Silva and Zanon, 2011:

In search of a new perspective, it is understood that improving the quality of chemistry teaching involves defining a teaching methodology that favors contextualization as one of the ways of acquiring data from reality, providing an opportunity for critical reflection on the world and cognitive development, through their active, creative and constructive involvement with the content covered in the classroom.

In this sense, it can be seen that contextualization may be the main solution for improving chemistry learning. However, this is not the only aspect to be verified, as learning chemistry depends on many variables, such as: teacher, student, teaching resources, the socio-cultural environment in which they are inserted, and the way in which chemistry is approached. It is known that it is up to the teacher to be an interlocutor in the production of knowledge, to manage the class, promoting the production of knowledge, because to be a teacher is to educate (CONCEIÇÂO; BONFÂ, 2013).

Based on this, it is worth highlighting that the active participation of the student in this continuous and renewing process is of the utmost importance for teaching and learning, because starting from pre-molded knowledge, the educator will seek to mediate the construction of scientific knowledge through contextualization with the students' universe and its interconnection with the school universe (ANDRADE, 2012).

3.2. EXPERIMENTATION AND MOTIVATION

The teacher's mission is to transform society, because he or she is the main character in education. It is the only route to social integration for everyone, and the only way out of poverty for the poorest section of the population (SAVIANI, 2000). Therefore, all the laws, all the books, all the buildings, all the computers and all the government funding are useless if the teacher is not present in the classroom, whole, motivated, well-trained and aware (VEIGA, 2000).

The picture presented by public schools in relation to the classes taught by chemistry teachers is discouraging. It is recognized that there is a need to reformulate the teaching of chemistry in schools, since experimental activities are capable of providing students with better knowledge (AMARAL, 1996). Experimentation played an essential role in consolidating the sciences from the 18th century onwards. At that time, there was a break with current research practices, which still considered a close relationship between nature and man, where it occupied a privileged place in the proposition of a scientific methodology summarized by the regularization of procedures (QUEIROZ, 2004). However, chemistry is an experimental science; it is therefore very difficult to learn it without carrying out practical (laboratory) activities. These activities can include demonstrations by the teacher, experiments to confirm information already given, the interpretation of which leads to the development of concepts, among others (MALDANER, 1999).

Students' lack of interest in studying chemistry is generally due to the lack of experimental activities that can link theory and practice. Teaching professionals, in turn, claim that this problem is due to the lack of laboratories or equipment to enable practical lessons to be carried out (QUEIROZ, 2004).

QUEIROZ teaches that in the particular case of high school chemistry, knowledge should be part of a functional structure that allows us to predict or explain the behavior of material systems, both in situations of theoretical study and of experimental facts that occur in the laboratory or in everyday life. This knowledge structure must be based on simple principles and models, which can be applied as widely as possible, in order to be able to explain a wide variety of experimental events with few satisfactory theoretical schemes. This is possible if teaching is conducted in such a way that students learn principles, based on very well-developed concepts, without letting them get lost in the particular study of isolated facts. In this way, the facts will be rationally grouped and it will be easier for the student to integrate and retrieve the information because it is related to fundamental principles that serve as a reference.

Experimentation can be inductive, in which case students can control variables and discover or rediscover functional relationships between them, and it can also be deductive when they have the opportunity to test what is said in the theory, but the use of these well-planned activities greatly facilitates the understanding of the production of knowledge in chemistry, It can include demonstrations by the teacher, experiments to confirm information already given, the interpretation of which leads to the elaboration of concepts, among others. These activities are important in forming links between spontaneous conceptions and scientific concepts, providing students with opportunities to confirm their ideas or restructure them (GIORDAN, 1999).

Based on this context, one of the challenges for teachers who teach Science, Chemistry, Physics and Biology is to put the theory they have studied into practice, given the conditions offered by the school in terms of laboratories. It has already been proven that practical lessons help to retain the content studied, as well as preparing students for the construction of knowledge, understanding and its development (SANTANA, et al, 2010).

Science teachers are well aware of the fact that experimentation arouses a strong interest among students at different levels of schooling. In their statements, students also often attribute to experimentation a motivating, playful character that is essentially linked to the senses. On the other hand, it is not uncommon for teachers to say that experimentation increases learning capacity, as it works as a way of involving the student in the topics at hand (GIORDAN, 1999).

According to Izquierdo et al. (1999), experimentation at school can have various functions, such as illustrating a principle, developing practical activities, testing hypotheses or as an investigation. However, the latter is what helps students learn the most.

Guimaraes (2009) states that experimentation can be an efficient strategy for creating real problems that allow contextualization and stimulate investigative questions.

Beltran and Ciscato (1991), in describing the history and main problems of chemistry teaching, point to the lack of well-planned experimental activities that allow students to experience an investigative situation in which they can learn how chemical knowledge is constructed.

According to Silva and Zanon (2000), the central aspects relating to the problems of experimentation in chemistry teaching concern the lack of teacher training, especially the lack of clarity about the role of experimentation in student learning.

The subject of motivation in relation to learning is always in the spotlight in school environments, pushing teachers to excel or making them retreat, even giving up in the most complex cases. However, it plays a very important role in the results that teachers and students are aiming for. However, it is a great challenge for education to develop students' skills, discovery attitudes, critical sense and creativity, leading them to understand and transform the world around them (MORAES; VARELA, 2007).

Considering that motivating means providing a reason for learning, i.e. stimulating the will to learn, it is understood that, in educational work, individual differences must be respected, as the same incentives do not achieve the same result for students of different ages and cultural backgrounds. In short, learning requires motivation (ANDRADE, 2012).

According to Bzuneck (2000), "motivation, or motive, is what moves a person, puts them into action or makes them change course".

Motivation can be understood as a process and, as such, it is what prompts or incites a behavior, which sustains a progressive activity, which channels this activity towards a given direction (BALANCHO; COELHO, 1996).

In the teaching-learning process, it is believed that motivation must be present at all times. It is up to the teacher to facilitate the construction of the training process, influencing the student in the development of motivation and learning (ANDRADE, 2012).

According to Piaget (1977), knowledge is formed through the interaction of program content with reality, not through copying but through assimilation. Educators must be prepared to develop meaningful learning, using a variety of techniques and contextualized methods.

However, it is still seen today that the teaching of chemistry in most schools is done through traditional teaching techniques (theoretical) and memorization of formulas, in other words, an abstract teaching of the content. This consequently creates a lack of motivation and rejection on the part of students, who see chemistry teaching as a dry and uninteresting area of knowledge, and sometimes unnecessary to their daily lives (ANDRADE, 2012).

In order for chemistry teaching to become motivating, it is necessary to use innovative teaching strategies, such as contextualized experimental classes, extra classes, interdisciplinarity, seminars and meetings. Making students curious and able to identify where the teaching of chemistry is related to their daily lives, thus enabling them to learn and develop the knowledge they are working on (ANDRADE, 2012).

3.3. computers and chemistry teaching

We are living at a time when the world is facing a revolution in communications between peoples through communication and information technologies (ICTs). Among the various sectors of society, education is one of the areas that is being affected by this technology (FERREIRA, 1998). As a result, we are also facing a time of review of school education, its role and its scope. Along with this comes the challenge of building a professional profile for teachers based on their work in the classroom, but which extends to the development of the school's educational project, the production, systematization and socialization of pedagogical knowledge and participation in discussions in the educational community (BENITE, A.; BENITE, C., 2008).

The National Curriculum Parameters emphasize that teachers need to be able to get to know their students, to adapt the teaching-learning process, to develop activities that enable the use of new communication technologies and information. In short, they must seek quality teaching that is capable of forming critical citizens. Promoting activities that can stimulate and help students understand concepts, such as questioning, debate, investigation, group work and the use of technology. In this way, students come to understand science as a historical construction, without taking into account teaching based on memorizing definitions and classifications (BRASIL, 1998).

Communication technologies, as well as being conduits of information, enable new ways of organizing human experience, with multiple effects, particularly on cognition and human action in the environment and on oneself. The use of information market products - magazines, newspapers, books, CD-ROMs, radio and television programs, homepages, websites, electronic mail - not only makes new forms of communication possible, but also generates new ways of producing knowledge. A few years ago, there was no possibility of online communication between physically distant people, or of sharing images instantaneously in various parts of the world, nor was it possible to conceive that a person could learn with a machine as their interlocutor, as is the case with computer-mediated learning (BRASIL, 1998).

According to Brasil (1998):

The computer, in particular, enables new ways of working, making it possible to create learning environments in which students can research, make anticipations and simulations, confirm previous ideas, experiment, create solutions and construct new forms of mental representation. It also allows interaction with other individuals and communities, using interactive communication systems: computer networks.

Starting from this context, creating a relationship between the subjects of Chemistry and Computer Science can "generate an environment favorable to teamwork, creating integration and promoting the manifestation of creativity through the performance of activities in small groups" (BENITE, A.; BENITE, C., 2008), or even individually.

The possibility of the teacher appropriating these technologies and integrating them with the chemistry teaching-learning environment can also generate more dynamic chemistry teaching that is closer to the constant transformations that society has experienced, helping to reduce the distance that separates basic education from modern tools for producing and disseminating knowledge (BENITE, A. and BENITE, C., 2008).

Observing all this, teachers can create this relationship between chemistry and computer science through a range of free and paid software that "allows interactivity between user and knowledge, which can enable or facilitate meaningful learning of chemical content" (ANDRADE, 2012).

It is believed that when using computers in the classroom, teachers must prepare themselves, devising teaching strategies that attract students to the content being taught, so that they can contribute to the construction of knowledge. And that the use of software or multimedia programs is not just the sequential visualization of texts, figures or videos, but that there is active student participation. In other words, it is necessary for software or multimedia programs to offer strategies for interacting with users, for example, making them think and propose solutions to a problem (MEDEIROS, 2008).

The school must integrate ICTs because they are already present and influential in all spheres of social life. As a result, the possibility of the teacher appropriating these technologies and integrating them into the chemistry teaching-learning environment can generate more dynamic chemistry teaching that is closer to the constant transformations that society has experienced, helping to reduce the distance that separates basic education from modern tools for producing and disseminating knowledge (BENITE, A.; BENITE, C., 2008).

In certain areas, such as chemistry, teaching practice can be favored by experimentation as an instructional tool. The learning of many chemical concepts is favored when there is an experimental approach. This aspect should be taken advantage of as a facilitator of teacher-student interaction, although it poses material difficulties that make experimentation prohibitive in schools with few resources. It is important to evaluate how the use of computers can contribute to the educational process since, in an attempt to contextualize theory and practice in the teaching of chemistry, the use of computer resources in classes can represent a viable alternative [1]. One type of computer program that can be used for didactic purposes is simulation programs, which make it possible to highlight specific aspects of the content covered and guide decision-making in experiments, a situation that greatly favors the understanding of chemical concepts (TORETTI, 2000).

3.3.1. The *Yenka software*

Yenka (Figure 1) is an educational modeling tool created in 2006 by *Crocodile Clips*. It has a series of simulations and models made specifically for educational institutions. It is a series of programs that can be sold together or separately, covering Mathematics content (*Yenka Mathematics*), which allows you to experiment with statistics, probability, coordinates and 3D drawings; Technology (*Yenka Technology*), a complete electronic design simulator, simulating and designing circuits with more than 150 types of components, as well as working in 3D format, among many other features in the area; *Computing* (*Yenka Computing*), with innovative tools for introducing programming through the control of animated characters, also in 3D; and Physics and

Chemistry (*Yenka Science*), which will be more detailed in the next section.

Figure 1 - Initial image of the *Yenka* software, with *Yenka Science* highlighted.

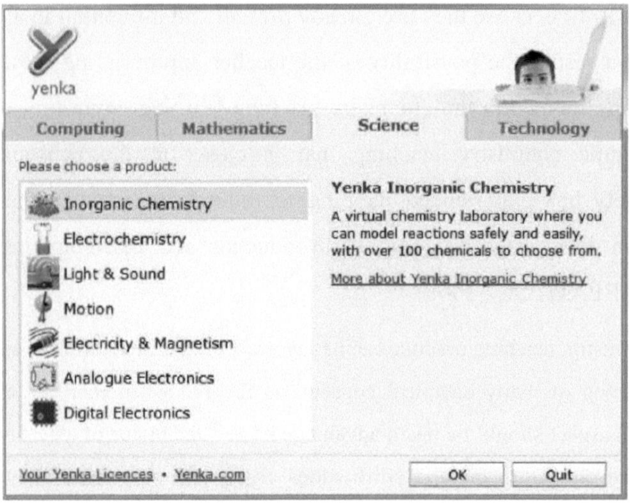

Yenka is available in English, but because it is so illustrative, this does not prevent it from being used easily by students. It requires a license, which varies according to the type you want. There is the *Teacher License*, which allows the teacher to use the program with up to 40 students, and the *School Site License*, which is more expensive but allows an unlimited number of users to use its *online* resources.

Yenka products are sold separately with prices ranging from \$225 to \$1500 dollars. Users can choose to buy some packages at a discount. In the case of *Yenka Sciene*, for chemistry and physics lessons, the price is \$450, with a teacher's license. If the user only chooses Yenka Chemistry, only for chemistry lessons, their license will cost \$225 dollars.

Crocodile Clips also offers free versions of *Yenka*. There is a version in which the user can use the *software* at home, and use all the content the *software* offers. This version does not allow use in schools, but works on weekends and after 3pm on working days. There is also a trial version that can be used in schools for 15 days. The latter was the version used in this work.

3.3.1.1. Yenka Science

Yenka Science (Figure 2) is a highly interactive virtual laboratory that allows to easily model Physics and Chemistry experiments in a safe way. *Yenka Science* comes with *Yenka Physics* and *Yenka Chemistry*. They simulate experiments such as: designing optical and circuit systems, modeling chemical reactions using more than 100 different chemicals, investigating propagating waves or accelerating masses, and the user can follow the progress of the experiment using graphing tools. There is also a wide range of pre-prepared lessons and a set of training videos to help the user get to grips with the *software*. The program covers the areas of Light and Sound, Force and Motion, Electricity and Magnetism, Inorganic Chemistry, Physical Chemistry and Electrochemistry.

Figure 2 - *Yenka Science* interface.

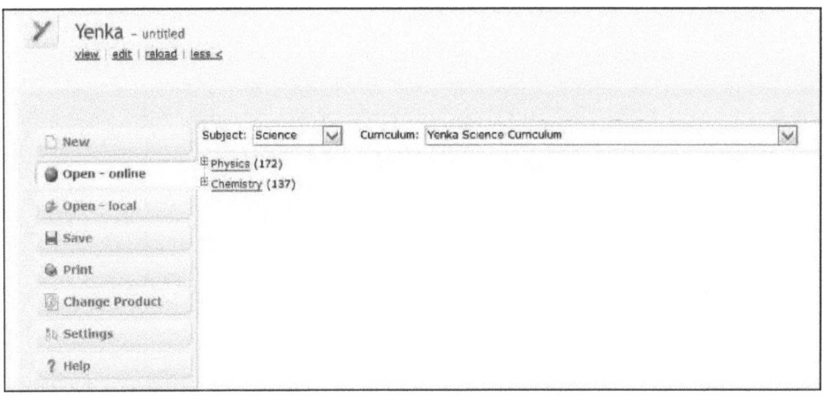

3.3.1.2. Yenka Chemistry

The *Yenka Chemistry software* is a virtual laboratory where experiments and reactions can be carried out easily and safely, with the facility to observe reaction results and various properties, just like a conventional chemistry laboratory. You can also find several ready-made models, separated by content, both *online* (Figure 3) and *offline* (Figure 4).

Figure 3 - Ready-made templates, separated by content, *online*.

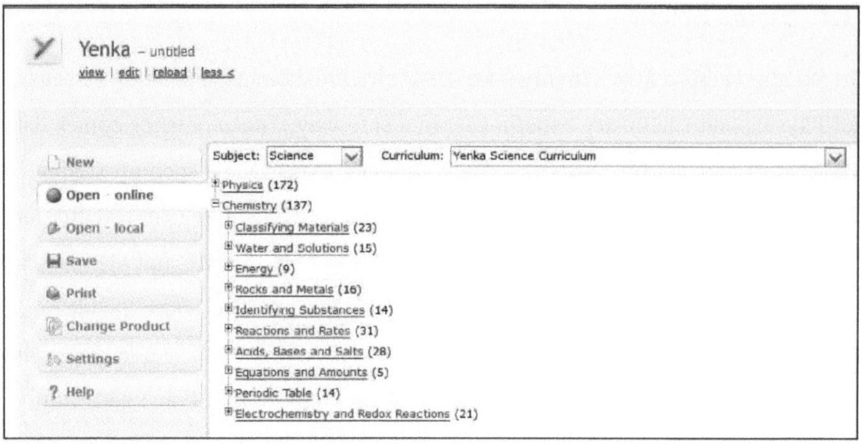

Figure 4 - Ready-made templates, separated by content, *offline*.

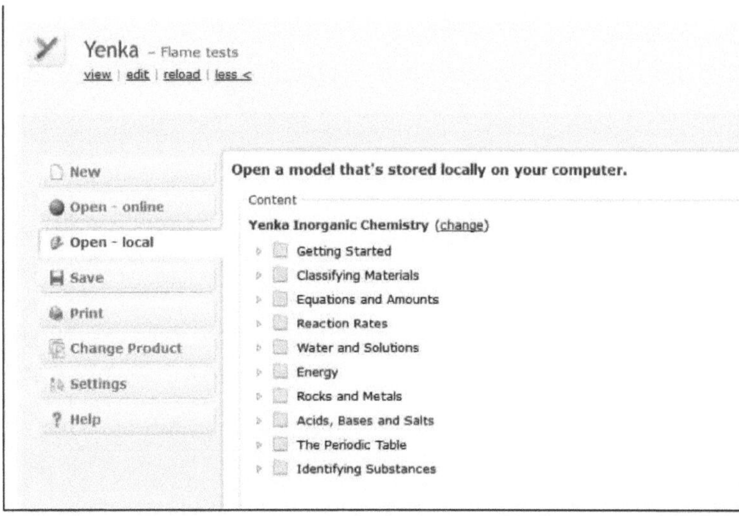

In the work area there are toolbars on which they are distributed (Figure 5), where various reagents, glassware, equipment and instruments that are used in a chemistry laboratory are available.

Figure 5 - Presentation and location of the toolbars on the desktop.

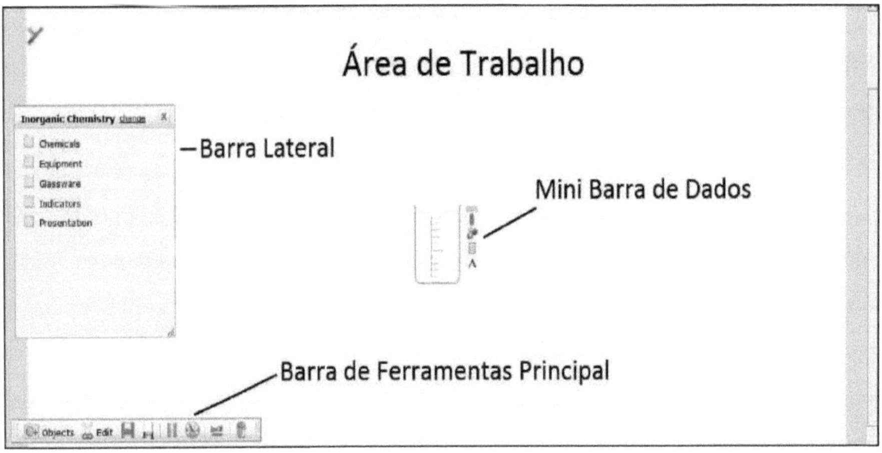

In the main toolbar you can edit (Figure 6), save, pause, speed up or slow down, among others (Figure 7).

Figure 6 - Functions of the *Edit* button on the main toolbar.

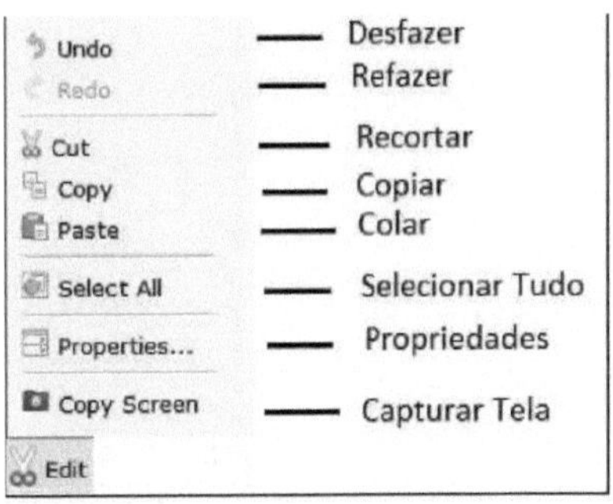

Figure 7 - Main toolbar.

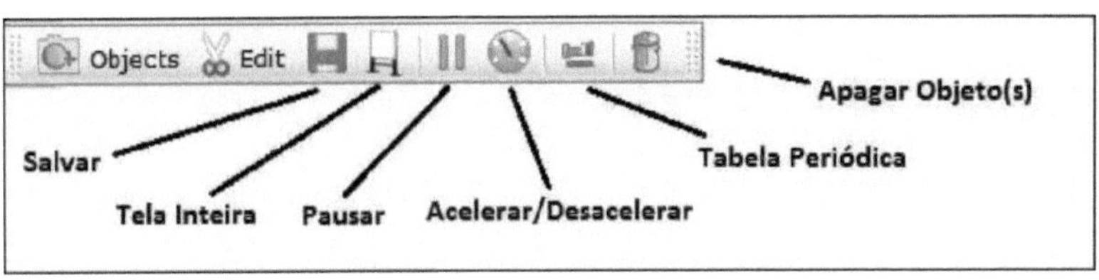

The sidebar can be activated via the *Objects* button on the main toolbar. With the sidebar you can use the materials and reagents that will be used in the experiment.

It is worth noting that the content of the sidebar varies according to the topic that the user will choose to work on in the initial part of the program, as can be seen in Figure

1. For example, the user can choose to work with materials suitable for the subject of *electrochemistry* by selecting the *Electrochemistry* option. For this work, the first option was selected, *Inorganic Chemistry*, which will be detailed below.

In *Chemicals*, you can find all the reagents you need for the experiment. Under *Equipment*, you can find laboratory equipment ranging from a simple stopper to a pH meter. Under *Glassware* you will find a variety of *glassware*. The *Indicators* folder allows you to use acid-base indicators. This can be done using graphs, pH papers or even solutions. Finally, the *Presentation* folder. This has resources that allow you to better identify certain materials, insert tables of texts, graphs, numbers, photos, animations and more.

The mini data bar is displayed when the mouse pointer is hovered over any glassware in the work area, whether or not it has anything inside. It will appear next to any piece of glassware, allowing you to obtain a range of information about the reaction, clean the container and name it (Figure 8).

Figure 8 - Glass displaying the mini data bar.

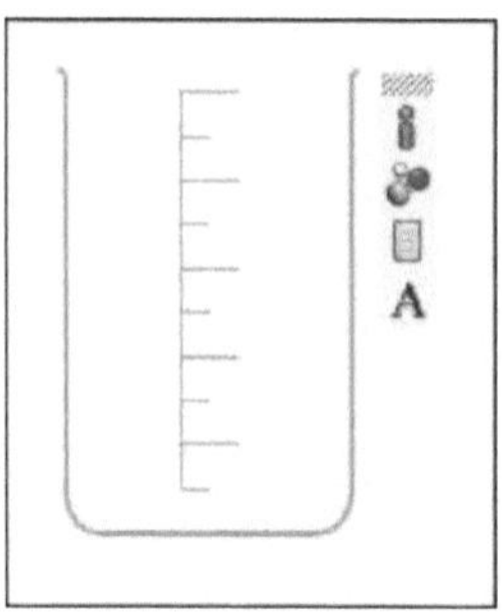

3.4. THE SUBJECT OF ACID RAIN

Rainfall in regions where the air is considered unpolluted usually has a pH of around 5.6. It has therefore been agreed to call acid rain rain that whose pH is lower than 5.6. Acid rain, defined in this way, can occur due to both natural processes and human activities.

Considering that acid rain is formed by natural processes, pH values below 5.6 in rainwater are generally the result of the presence of nitric acid, $HNO_{3((aq))}$, and sulphuric acid, $H_2SO_{4((aq))}$.

Nitric acid is formed naturally during storms by the electrical energy provided by lightning.

Sulphuric acid is formed mainly from sulphur dioxide, $SO_{2(g)}$, which is released directly into the atmosphere by volcanic eruptions.

Thus, in South America, rainfall with an average pH of 4.7 is common both in urban and industrialized areas and in sparsely inhabited regions.

Considering human activities, acid rain is formed from the burning of fossil fuels such as coal and oil, used to power factories, industries, thermoelectric power stations and transport vehicles, which dump thousands of tons of polluting gases into the atmosphere, which has increased the acidity of the rain.

It is estimated on a global scale that at least 50% of atmospheric sulphur dioxide comes from human activities. The irony is that rain with a high degree of acidity doesn't necessarily fall in the region that produced the polluting gases. For example, air currents moving from the ocean to the continent carry most of the annual 5 million tons of $SO_{2(g)}$ produced in coal-fired power stations in the British Isles to southwestern Sweden and southern Norway, where high acid rainfall occurs. The result is that, despite being countries with low emissions of polluting gases, Sweden and Norway are among the countries that suffer the most from the harmful effects of acid rain. In Sweden, acidification has compromised fishing in 2,500 lakes and it is estimated that another 6,500 have already been affected. Of Norway's 5,000 lakes, 900 are seriously affected and 1,750 have lost all their fish.

It's not just living beings that suffer from acid rain. It is believed that monuments such as the Parthenon in Athens (Greece) have deteriorated more in the last 25 years than in all the previous 2400 years.

The same phenomenon can be seen at the Taj Mahal in India and the Colosseum in Rome. This is because the solution of sulphuric acid, $H_2SO_{4((aq))}$, which falls as rain reacts with the marble, $CaCO_{3(s)}$, of the monuments, turning the surface into plaster, $CaSO_{4(s)}$, which is soft and subject to erosion.

Not to mention that the $H_2SO_{4(aq)}$ in the rain also corrodes the metallic iron present in the steel used in the structures of houses, buildings, bridges and vehicles, causing immense damage to people and nations (FONSECA, 2003).

The topic of acid rain is an important one to work on with students because, as well as being a viable topic for the National High School Exam, it is a way of promoting environmental education.

According to the National Curriculum Parameters (PCNEF), Environmental Education should be developed with the aim of helping students to build a global awareness of environmental issues (BRASIL, 1999).

"Education in Chemistry and the Environment" is more than a proposal to color the chemistry we work with in secondary and elementary schools with some fundamental concepts of Environmental Chemistry. It aims to make chemistry teaching effective in understanding the environmental problems that surround the local community and the globe itself. It is to prepare citizens to make decisions based on the chemical sciences. It means reviewing prejudices and proposing effective solutions that are also anchored in this science (NETO, 2006 *apud* CORREIA, 2009).

As already mentioned, the topic used in this work was acid rain. However, a number of thematic or conceptual lessons can be taught with the help of *Yenka Chemistry*: alcohol content in petrol, reactions involved in the formation of a breathalyzer, concentration of solutions, reactivity of alkali metals, stoichiometric calculation, gases, enthalpy and speed of chemical reactions, density of compounds etc.

4. METHODOLOGY

4.1. DIDACTIC STRATEGY

This work sought information on the students' knowledge and learning of the subject and the method used in class, using the *Yenka* virtual laboratory, with the subject of acid rain as the content. The students were given an introduction to the *Yenka software*, where they were taught some basic processes for using the program, as well as how to purchase the product.

The topic chosen gave rise to two investigation scripts. The first script (Appendix A) was based on the formation of acid rain, using an experimental kit already available in the program, in which the student, on completing the procedures in the script, had to fill in two tables. It's worth pointing out that the tables to be filled in required the students to observe the components in solid, liquid and gaseous states, as well as the reactions that took place there. It also included pH and substances in aqueous solutions.

The students had to observe the experiment and then fill in the tables according to their observations (Figure 9). The students had to be able to identify the formation of sulphur dioxide, via the information box, when it comes into contact with water, turning it acidic, and associate the virtual process with the real chemical phenomenon. After filling in the tables, some questions were asked about the phenomena observed.

Figure 9 - Students investigating the process of acid rain formation.

The second script (Appendix B) aimed to help students understand the damage that acid rain can cause. Based on this, the script showed the reaction of sulphuric acid with a limestone (calcium carbonate). The students had to visualize the limestone being corroded by the acid, simulating the real phenomenon, as well as the formation of carbon dioxide gas through the bubbles that were released from the beaker. At the end of the observations, a table had to be filled in which contained information such as the formation of gases, solids contained or formed and substances in aqueous solution. Due to the time limit offered by the school, it was not possible to apply this script in full. The students only filled in the table, without having answered the questions at the end of the script. However, this did not affect the final results of the analysis.

The data and results obtained from the scripts and questionnaires were analyzed and are presented in the next chapter.

4.2. ACTION STRATEGIES

This work was carried out at the Presidente Mèdici State School of Elementary and Secondary Education (Figure 10) in the city of Joao Pessoa - PB in a 3rd year high school class with 34 students, who answered a questionnaire (Appendix C) after a lesson using the program. This class was chosen because the students had already studied some concepts of acids and bases in the previous year and because of the ease of access to the class.

Figure 10 - Presidente Mèdici State School. Location of the course completion work.

In the morning, the school works with integrated vocational education, where students, in addition to studying regularly, take vocational courses in IT or Hospitality, the latter of which was the class that took part in the lesson and answered the questionnaire. In the afternoon, there are regular high school classes, without integrated courses. Finally, in the evening, there are classes in the Youth and Adult Education (EJA) program.

The computer lab (Figure 11) had 18 computers, all with the Linux Educational operating system, but due to problems with the wireless connection network, it was not possible to install the *Yenka software* on all of them, and it was only possible to install it on 8 of them. As a result, the students had to work in trio or pair groups. Those who had difficulty following the procedures, due to the small number of computers that had the program properly installed, were able to observe the phenomena using the Datashow system installed there.

Figure 11 - Computer lab at the Presidente Mèdici State School.

5. RESULTS AND DISCUSSION

5.1. SCRIPT ANALYSIS

Analyzing the first script (Appendix A), which dealt with the formation of acid rain (Figure 12), it can be seen that the students, when filling in the tables, did not have much difficulty in visualizing the chemical transformations, using the information box that the program offered.

Figure 12 - Experiment related to script 01: formation of acid rain.

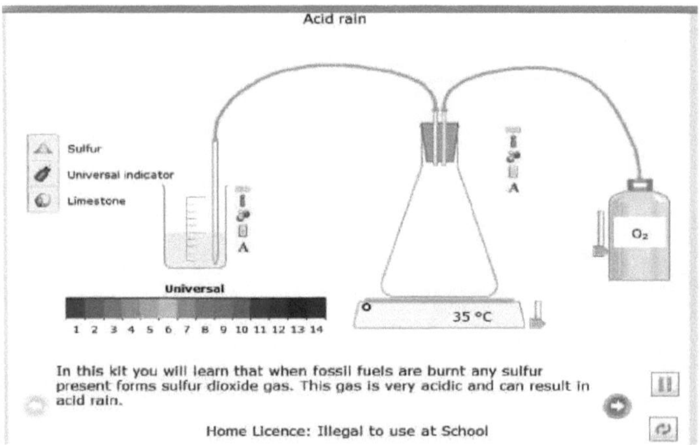

Graph 1 shows that most of the students managed to fill in the tables correctly, i.e. of the 34 students in the computer lab, 28 filled them in correctly, 5 filled them in partially correctly, with only one of them filled in incorrectly or even blank, and only 1 left all the tables blank.

Graph 1 - Graph of the number of students who filled in the tables in the script 0

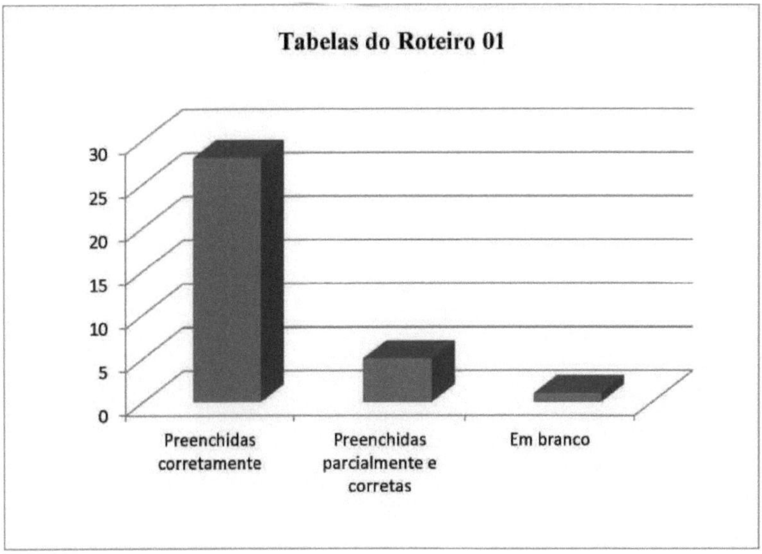

As for a small final questionnaire, already included in the script, we can see a relative variety of answers. Let's take a look:

Question 01

"When you added the universal indicator to the water, you noticed that it turned green. However, after connecting the oxygen gas cylinder and the electric heater, the water turned a very strong reddish color. Why did this happen?"

Of the 34 students, 21 answered correctly, 4 got it partially right, 1 got it wrong and 8 left it blank. Looking at Graph 2, it can be seen that most of these students answered the question correctly.

The fact is not that the question was simple, but that the students associated the virtual experiment with the theory. They noticed that there was a color change in the solution, which became increasingly acidic. This shows that the virtual laboratory helped them to answer the question.

The option "partially right" was added because the students answered the questions using expressions or words that were not appropriate, but which did not lose the meaning of their explanations. This option was used for all the questions.

Graph 2 - Analysis of the results of the first question of script 01.

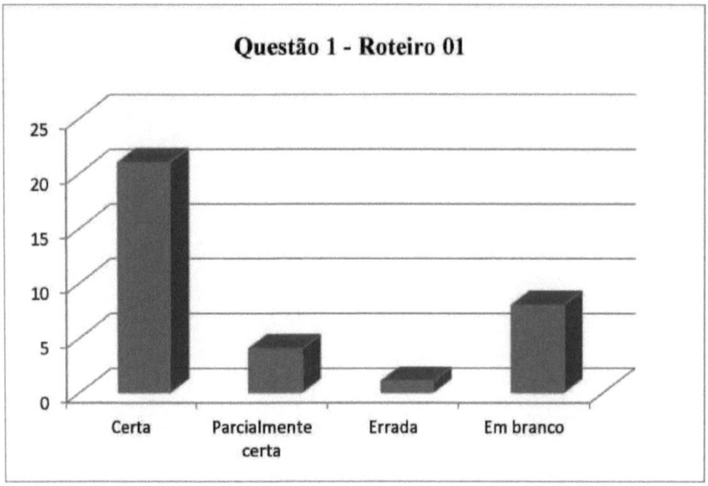

Question 2

"What is the main gas that causes acid rain?"

This second question, which is much simpler but no less important, had 16 correct answers, 5 partial answers, 8 errors and 5 blank answers, as can be seen in Graph 3. Partially correct answers were those who answered that sulphur was the cause, even though the question asked for the causing gas.

Looking at the graph for this question, it can be seen that most of the students got it right, showing that the method of using the virtual laboratory in chemistry classes was able to provide more meaningful and motivating learning.

Graph 3 - Analysis of the results of the second question of script 01.

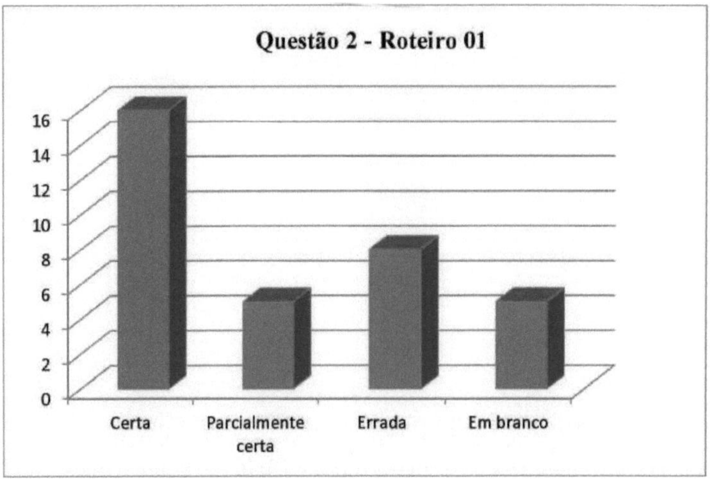

Question 3

"Based on what you've learned from the experiment, explain how acid rain is formed."

This last question demanded a little more of the student. As a result, there were more blank answers (Graph 4), although the number of correct answers was still higher. So there were 14 correct answers, 7 partial correct answers and 13 blank answers. There were no wrong answers.

The possible reasons for the high number of blank answers could have been the small number of computers available for the number of students, which could have made it difficult for those who only observed the experiment using the Datashow to understand. It could also be that, because it was the beginning of the school year, the students might not have remembered some of the content needed for the experiment, even though they had been taught the subject before they started.

Graph 4 - Analysis of the results of the third question of script 01.

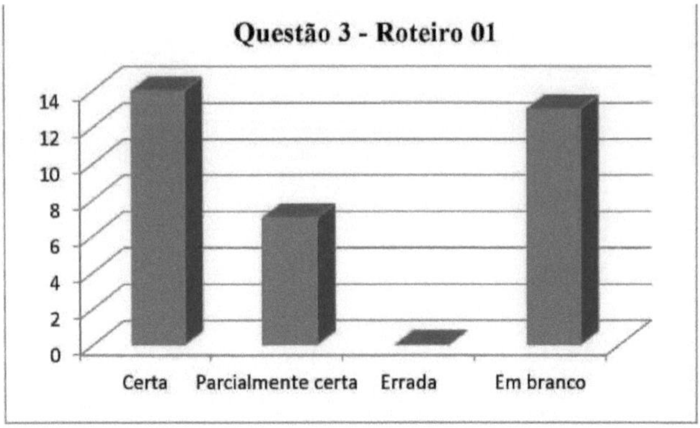

The second script (Appendix B) could not be applied in full due to the limited time offered by the school. The vast majority of students did not fill in the table in the second script, which could have jeopardized data processing. For this reason, it was necessary to dispense with the second script of this work on the effects that acid rain can cause.

5.2. ANALYSIS OF THE FINAL QUESTIONNAIRE

At the end of the lesson, a short questionnaire (Appendix C) was administered to evaluate the methodology applied to the students.

The following graph shows the students' answers to the first question in the questionnaire.

The graph clearly shows that the virtual laboratory helped the students in their learning. Of the 34 students who answered the questionnaire, 32 said yes, only one said no, and one left it blank.

Graph 5 - Question one of the final questionnaire.

We can highlight some of the justifications given in the questionnaire:

"It helped a lot, because in practice everything becomes easier."

"Because we learned a lot of things. If there hadn't been a virtual laboratory it wouldn't have been very good."

"It's more fun, as if we were in a laboratory and not in danger from the experiments."

Looking at the answers to the second question, it can be seen that if the students had only attended lectures on the subject of acid rain, most of them would not have fully understood the concepts involved. Of these, 25 answered *no*, 7 answered *yes*, and only 2 left the question blank.

Graph 6 - Survey of students' answers to the second question.

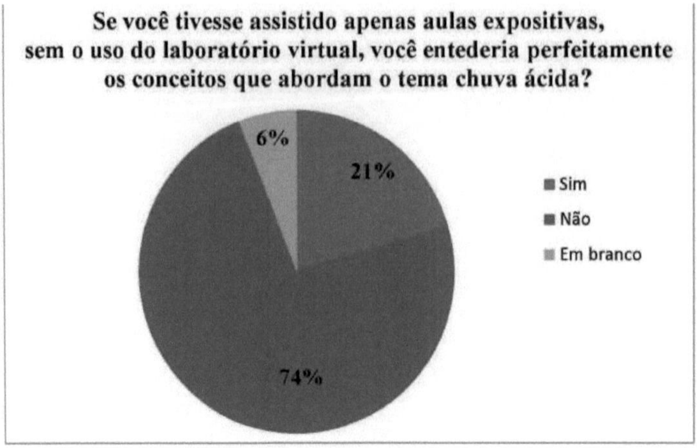

We can highlight a few comments:

"Because images showing the whole process make it easier to understand."

"Because it would be monotonous and with the virtual class it was a different way of learning chemistry."

The third question in the questionnaire (Graph 7) further strengthens the importance of the virtual laboratory in the students' teaching-learning process. Similar to the first question, 32 students answered that the use of *software*, such as *Yenka*, is important in chemistry lessons. Only one answered no, and one left the question blank.

Graph 7 - Survey of students' answers to the third question.

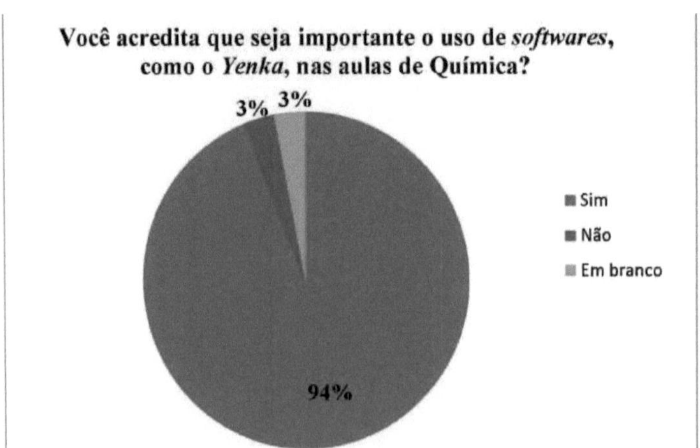

A few comments are worth highlighting:

"Because since we don't have a real laboratory, the virtual one is the only way to learn better."

"Because it makes it possible to experiment without risk."

"It would increase school performance and chemistry lessons would be more useful. "

There was also an optional fourth question, which asked for criticism or suggestions from the students. These included:

"The class was great. I hope to attend more classes like this."

"I found it very interesting and cool."

"The virtual lab should have an option to change the language. Since it's in English."

Other students also questioned something similar to the last comment. They questioned the fact that the program is in English. However, it is a program that is easy to manipulate because it has a lot of illustrations. This was due to the unlimited number of classes the school was able to offer. Despite this, the language did not interfere with the students' satisfaction when using *Yenka*, but they would certainly have mastered the program perfectly with a few more lessons.

Despite some difficulties in implementing the project, the results were quite significant. One of the difficulties was the installation, as all versions can only be installed with an e-mail registration. But for this to happen normally, it was necessary to have an internet connection. The school's computers used wireless internet. As a result, the signal didn't reach all the computers, which made it difficult to install the program on all of them. As a result, the minimum number of computers meant that all the students couldn't fully interact with the program. For this reason, the Datashow system was installed so that students who had difficulty observing the processes and reactions proposed in the script could have a better view and understanding of the lesson.

According to the chemistry teacher in charge of the class, in order for the school to buy the program, it is necessary to contact the State Education Department via a letter, which would undergo a strict analysis as to the feasibility of acquiring the program, and then the purchase would be approved or not.

6. CONCLUSION

The *Yenka software*, as presented in this paper, is a virtual laboratory that has significant potential for teaching chemistry, and is a viable alternative for schools that don't have the resources to set up science laboratories.

The importance of using them in computer rooms, which have lower installation and maintenance costs than science laboratories, is obvious.

It was also seen that the students assimilated the information on acid rain in a more meaningful way, compared to a traditional lesson, which would make the phenomena somewhat abstract for them to understand.

In addition, the opinion poll, based on the final questionnaires answered by the students, points to the didactic and instructional contribution of this *software*. From the students' testimonies, it is clear that they are very accepting of the program. This positive point can be highlighted as an agent that facilitates and motivates learning.

Based on this context, a new learning and interdisciplinary environment can be created, combining proposals included in official documents for teaching chemistry with innovative teaching practices, resulting in more effective and interactive learning.

7. BIBLIOGRAPHICAL REFERENCES

ALMEIDA, E. C. S. et al. Contextualizing chemistry teaching: motivating high school students. In: ENCONTRO DE EXTENSÂO, 10., 2008, Joao Pessoa. **Electronic proceedings**... Joao Pessoa: UFPB, 2008. Available at: <http://www.prac.ufpb.br/anais/xenex_xienid/x_enex/ANAIS/Area4/4CCENDQPEX 01.pdf>. Accessed on: 18 Dec. 2016.

AMARAL, L. **Trabalhos prâticos de quimica**. Sao Paulo, 1996.

ANDRADE, R. W. N. **Utilization of the *software Crocodile Chemistry* 605 - virtual chemistry laboratory, as a tool for teaching chemistry.** Joao Pessoa: IFPB, 2012, 67p. Course Conclusion Paper, Federal Institute of Education, Science and Technology of Paraiba.

BALANCHO, M. J. S.; COELHO, F. M. **Motivar os alunos, criatividade na relaçâo pedagògica: conceitos e prâticas**. 2. ed. Porto, Portugal: Texto, 1996.

BELTRAN, N.O.; CISCATO, C. A. **Quimica**. Coleçao Magistério de 2° Grau. Sao Paulo: Cortez, 1991.

BENITE, A. M. C.; BENITE, C. R. M. The computer in chemistry teaching: impressions versus reality. A focus on public schools in the Baixada Fluminense, **Ensaio: Pesquisa em Educaçâo em Ciências**, Belo Horizonte, v. 10, n. 2, p. 303319, 2008.

BRAZIL, MEC. **As Novas Diretrizes Curriculares que Mudam o Ensino Médio Brasileiro**, Brasilia: Secretaria de Educaçao Fundamental/MEC, 1999.

. **National Curriculum Parameters - Third and Fourth Cycles of Primary Education**. Brasilia: Secretaria de Educaçao Fundamental/MEC, 1998. .

National Curriculum Parameters for Primary Education. Brasilia: Secretaria de Educaçao Fundamental/MEC, 1999.

BZUNECK, J. A. Teachers' self-efficacy beliefs. In: F.F. Sisto, G. de Oliveira, & L. D. T. Fini (Orgs.). **Psychology readings for teacher training** . Petrópolis, Rio de Janeiro:

Vozes, 2000.

CONCEIÇÂO, E. B. O.; BONFÂ, M. B. Difficulties in teaching and learning chemistry in the 1st year of secondary school: a case study at the Cora Coralina State School of Elementary and Secondary Education. **Revista Saberes**, Pimenta Bueno, v. 1, n. 2, Jul./Dec. 2013. Available at: <http://www.athenaseducacional.com.br/fap/revista/images/segunda_edicao/dificulda teaching_and_learning_in_chemistry.pdf>. Accessed on: 20 Jan. 2016.

CORREIA, P.; **Acid rain in the conception of high school students in the municipality of Alagoa Nova - PB**. Campina Grande: UEPB, 2009, 63p. Guided Academic Work, State University of Paraiba.

FERREIRA, V. F. Interactive technologies in teaching. **Quimica Nova**, Sao Paulo, v. 21, n. 6, p. 780-786, Nov./Dec. 1998.

FONSECA, M. R. M. Acid rain pollution. ln: FONSECA, M. R. M. **Interatividade quimica: cidadania, participçâo e transformçâo, ensino mèdio. Single** volume. Sao Paulo: FTD, 2003, p. 410-411.

GIORDAN, M. O papel da experimentaçao no ensino de ciências. **Quimica Nova na Escola**, Sao Paulo, v.10, n.10, p. 43-49, 1999.

GUIMARÂES, C. C. Experimentaçao no ensino de Quimica: caminhos e descaminhos rumo à aprendizagem significativa. **Quimica Nova na Escola**, Sao Paulo, v. 31, n. 3, p. 198-202, Aug. 2009.

HACK, J. R.; NEGRI, F. School and Technology: teacher training as a reference for change. **Ciência & Cogniçâo**, Rio de Janeiro, v. 15, p. 89-99, 2010.

IZQUIERDO, M.; SANMARTi, N. and ESPINET, M. Fundamentación y diseno de las prâcticas escolares de ciencias experimentales. **Enseftanza de las Ciencias**, v. 17, n. 1, p. 45-60, 1999.

MALDANER, O. A. Research as a perspective for the continuing education of chemistry teachers. **Quimica Nova**. Sao Paulo, v. 22, n. 2, p. 289-292, 1999.

MEDEIROS, M. A. A informàtica no ensino de quimica: análisis de um software para o ensino de tabela periódica. In: ENCONTRO NACIONAL DE ENSINO DE QUÎMICA, 14., 2008, Curitiba. **Electronic abstracts**. Curitiba: UFPR, 2008. Available at:< http://www.quimica.ufpr.br/eduquim/eneq2008/resumos/R0749- 2.pdf >. Accessed on: 22 Oct. 2016

MICHEL, R.; SANTOS, F. M. T.; GRECA, I. M. R. An internet search for tools for high school chemistry education. **Quimica Nova na Escola**, Sao Paulo, n. 19, p. 3-7, May 2004. Education in Chemistry and Multimedia section.

MORAES C. R.; VARELA S.; Student motivation during the teaching-learning process. **Revista Eletrônica de Educaçâo**, Sao Carlos, year 1, n. 1, Aug./Dec. 2007. Available at:

<http://web.unifil.br/docs/revista_eletronica/educacao/Artigo_06.pdf>. Accessed on: 20 Dec. 2016.

PIAGET on Piaget: The Epistemology of Jean Piaget. Directed by Claude Goretta. New Haven: Yale University Media Design Studio, 1977. Film available at: <http://www.youtube.com/watch?v=I1JWr4G8YLM>. Accessed on 26 Dec. 2013.

QUEIROZ, S. L. **From doing to understanding science: reflections on the learning of scientific initiation students in chemistry**. Ciência & Educaçao, Bauru, v. 10, n. 1, 2004.

RIBEIRO, C. K. G. et al. Use of educational software as tools to support the teaching of chemistry. In: JORNADA DE ENSINO, PESQUISA E EXTENSÂO, 9., 2009. Goiana. **Electronic abstracts**... Goiana: UFRPE, 2009. Available at: <http://www.eventosufrpe.com.br/jepex2009/cd/resumos/R1132-2.pdf>. Accessed on: October 17, 2016.

SANTANA, S. L. C. et al. **Sugestâo para Planejamento de Atividades Experimentais**. 2010. (Development of didactic or instructional material - Booklet: Sugestao para Planejamento de Atividades Experimentais). Available at: < http://w3.ufsm.br/ppgecqv/Producao/atividades_experimentais.pdf>. Accessed on 18

Jan. 2015.

SAVIANI, O. **Pedagogia histórico-critica: primeiras aproximçôes**. 7 ed. Campinas, SP: Autores associados, 2000.

SILVA, L. H. A.; ZANON, L. B. Experimentation in science teaching. In: SCHNETZLER, R.P.; ARAGÂO, R. M. R. **Ensino de Ciências: fundamentos e abordagens**. Piracicaba: CAPES/UNIMEP, p.120-153, 2000.

TEIXEIRA, A.C.; BRANDÂO, E.J.R. Software educacional: o difícil começo. **Novas Tecnologias na Educaçâo**, Porto Alegre, v. 1, n. 1, Feb. 2003.

TORETTI, G. A.; MARTELI, C.; ROSSI, A. V. Good results are possible in the difficult context: Teaching Chemistry, Informatics and public schools. In: ANNUAL Meeting of THE BRAZILIAN CHEMICAL SOCIETY, 23rd, 2000, Poços de Caldas. **Electronic abstracts**... Poços de Caldas: Brazilian Chemical Society, 2000. Available at:< http://www.sbq.org.br/ranteriores/23/resumos/0390-2/>. Accessed on 22 Oct. 2016.

VEIGA, I. P. A. et al. **Pedagogia universitària: a aula em foco**. Sao Paulo: Papirus, 2000. 247 p.

WARTHA, E. J.; SILVA, E. L.; BEJARANO, N. R. Cotidiano e contextualização no ensino de Quimica. **Quimica Nova na Escola**, Sao Paulo, v. 35, n. 2, p. 84-91, May 2013. Featured Scientific Concepts section.

YENKA: educational modeling. Version 3.4.1. Edinburgh: Scotland, United Kingdom. Available at:< http://www.yenka.com/en/Download_the_Yenka_installer/>.

Accessed on: 15 Jun. 2015.

8. APPENDICES

APPENDIX A

Script 01

Acid rain formation

Acid rain is caused by gases that are produced when fossil fuels, such as gasoline, are burned. One of these gases is sulphur dioxide (SO_2). This gas, when in contact with rainwater, reacts with the water, turning it acidic. When this acidified water precipitates in the form of rain, it is called acid rain. The aim of this first experiment is to simulate the process of acid rain formation.

1. Find the ready-made diagram in *Science > Open - local > Acids, Bases and Salts > Acid rain.* You should see something like the figure below:

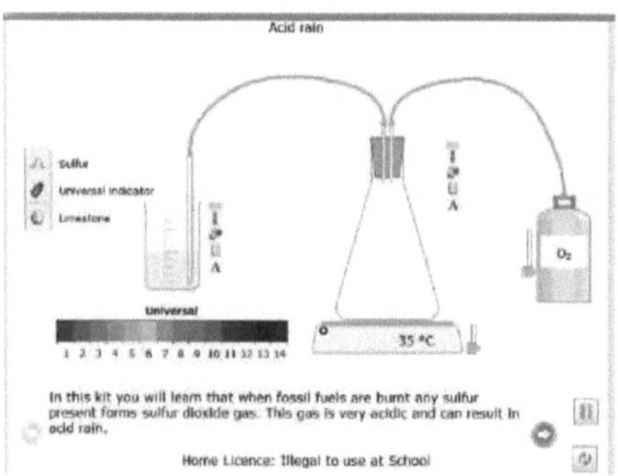

2. First, pause the process;

3. Add the sulfur powder to the enrlenmeyer flask. Then add the

universal indicator to water. Observe the color and compare it with the pH table. Write down the pH value. Open and maximize the *information box*. Don't forget to write down your observations, both in relation to what's in the beaker and the flask. Set it to *symbolic* mode (figure below). This way you can see more details;

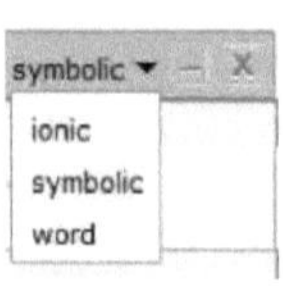

4. Turn on the oxygen gas cylinder using the button on the side of the cylinder. To release the gas, simply slide the button upwards;

5. Turn on the electric heater;

6. Disable the pause button and observe what happens to all the components of the experiment, always watching the *infrared box*;

7. Based on your observations and with the help of the information box, complete the tables below.

TABLE OF OBSERVATIONS - EXPERIMENT 01 - BECKER			
Gases	**Liquid(s)**	**Substance(s) in aqueous solution**	**PH**
Main chemical reaction			

TABLE OF OBSERVATIONS - EXPERIMENT 01			**- ERLENMEYER**
Solid(s)	**Gases formed**	**Liquid(s)**	**Substance(s) in aqueous solution**
Main chemical reaction			

8. When you added the universal indicator to the water, you noticed that it turned green. However, after connecting the oxygen gas cylinder and the electric heater, the water took on a very strong reddish color. Why did this happen?

9. What is the main gas that causes acid rain?

10. Based on what you have learned from the experiment, explain how acid rain is formed.

APPENDIX B

Route 02

Acid rain and its effects

Acid rain causes a lot of damage to buildings and materials, especially if they are made of limestone (calcium carbonate). This second script aims to help students understand the effects of acid rain.

1. Open the *Yenka* program, choose the *Science* option *and* click on *new ;*

2. First, click on the pause button;

3. Take a 100 mL beaker, click on *Glassware > Standard > Beacker (100 mL)* and drag it onto the work area;

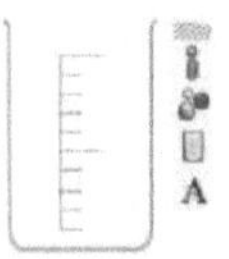

4. Take a limestone ($CaCO_3$), click on *Chemicals> Miscellaneous Lumps > > Limestone*, drag it into the beaker;

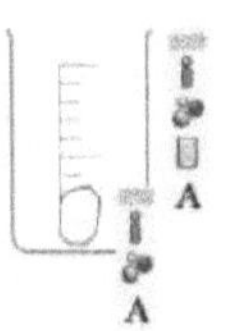

5. Now take 50 mL of *sulfuric* acid (H_2SO_4) and click on *Chemicals> Acids > Sulfuric acid.* Change the concentration to 0.5 M, so that the process is closer to the real thing. Now drag the acid into the beaker.

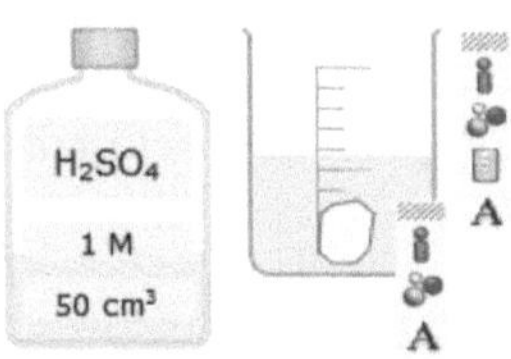

6. Write down the data of the substances in the table below:

DATA TABLE	
Acid =	**Salt =**
Concentration =	**Pasta =**
Volume =	

7. Now disable the pause button, watch what happens, and write down everything you see;

8. Open the *information box*, which is represented by the letter *i,* maximize it and put it in *symbolic* mode, and observe what is shown. Then complete the following table:

TABLE D]	**AND OBSERVATIONS - EXPERIMENT 02**	
Gas(es) formed	**Solid(s) contained and/or formed**	**Substance(s) in aqueous solution**

9. Set up the chemical reaction and, if necessary, balance it.

10. Why do monuments (statues and churches) made of limestone corrode when they come into contact with acid rain?

APPENDIX C

Questionnaire

The purpose of this questionnaire is to evaluate the methodology applied by the teacher. The accuracy of your answer is very important. Thank you.

1. Do you think the virtual lab helped you learn? Please justify your answer.

( ) Yes ( ) No

---

2. If you had only attended lectures, without using the virtual laboratory, would you have fully understood the concepts surrounding acid rain? Please justify your answer.

( ) Yes ( ) No

---

3. Do you think it's important to use software such as Yenka in chemistry lessons?

Please justify your answer.

(   ) Yes   (   ) No

4. Open space for suggestions or criticism (optional).

MIX
Papier aus verantwortungsvollen Quellen
Paper from responsible sources
FSC® C105338

Printed by Books on Demand GmbH, Norderstedt / Germany